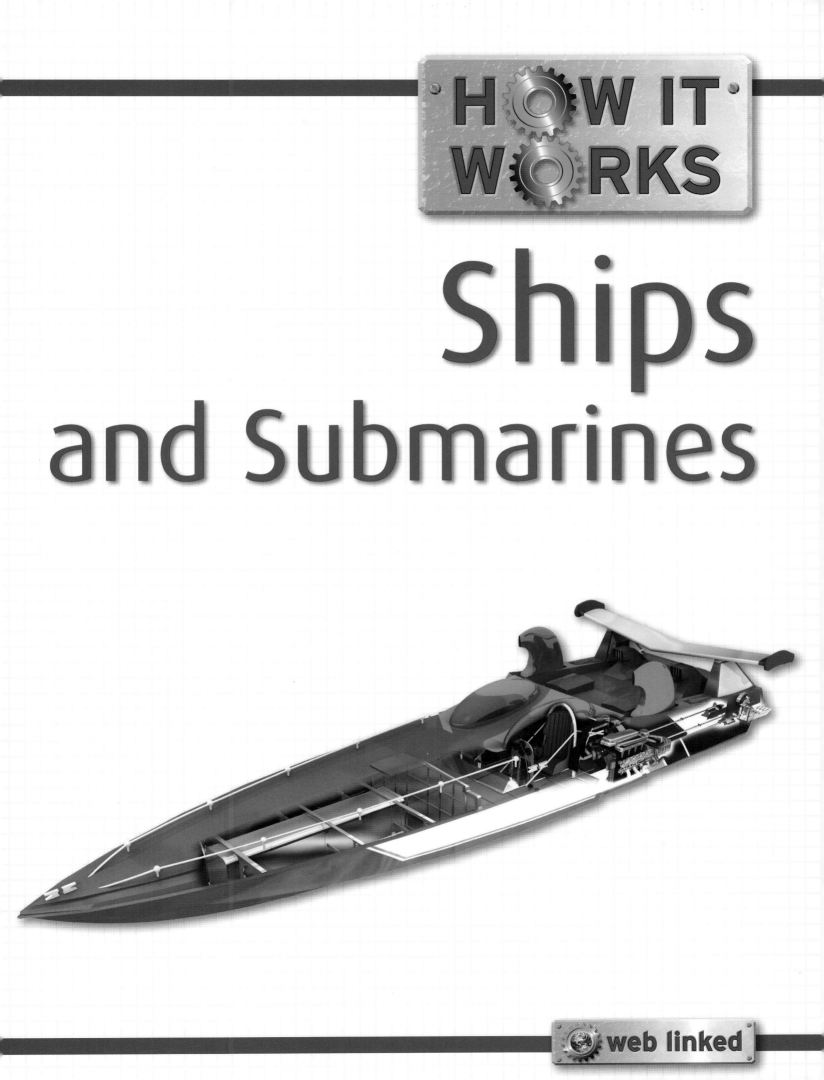

HOW IT WORKS

Ships
and Submarines

web linked

HOW IT WORKS

Ships
and Submarines

By Steve Parker

Illustrated by Alex Pang

MASON CREST PUBLISHERS INC.
370 Reed Road, Broomall, Pennsylvania 19008
(866)MCP-BOOK (toll free), www.masoncrest.com

First Printing
9 8 7 6 5 4 3 2 1

      Library of Congress Cataloging-in-Publication Data
Parker, Steve, 1952–
 Ships and submarines / by Steve Parker ; illustrated by
Alex Pang.
     p. cm. — (How it works)
 Originally published: Essex : Miles Kelly Pub. Ltd., c2009.
 Includes bibliographical references and index.
 ISBN 978-1-4222-1798-6
 Series ISBN (10 titles): 978-1-4222-1790-0
 1. Ships—Juvenile literature. 2. Submarines (Ships)—
Juvenile literature. I. Pang, Alex, ill. II. Title.
 VM150.P3485 2011
 623.82—dc22
               2010033635

Printed in the U.S.A.
───────────────
First published by Miles Kelly Publishing Ltd
Bardfield Centre, Great Bardfield, Essex, CM7 4SL
© 2009 Miles Kelly Publishing Ltd

Editorial Director: *Belinda Gallagher*
Art Director: *Jo Brewer*
Design Concept: *Simon Lee*
Volume Design: *Rocket Design*
Cover Designer: *Simon Lee*
Indexer: *Gill Lee*
Production Manager: *Elizabeth Brunwin*
Reprographics: *Stephan Davis, Ian Paulyn*
Consultants: *John and Sue Becklake*

Every effort has been made to acknowledge the source and
copyright holder of each picture. The publisher apologizes
for any unintentional errors or omissions.

ACKNOWLEDGMENTS

All panel artworks by Rocket Design
The publishers would like to thank the following
sources for the use of their photographs:
Corbis: 16 Andy Newman/epa; 18 Atlantide
Phototravel; 27 Paul A. Souders; 29 Lester Lefkowitz
Fotolia: 6 (b) Alexander Rochau; 7 (b) Snowshill;
9 Forgiss; 11 linous; 33 Aaron Kohr
Rex Features: 13 Neale Haynes; 15 Stuart Clarke;
21 Sipa Press; 23; 31 Sipa Press; 37 c.W. Disney/Everett
Photo Library: 7 (c) Bernard van Dierendonck
Science Photo Library: 34 Alexis Rosenfeld
All other photographs are from Miles Kelly Archives

WWW.FACTSFORPROJECTS.COM

Each top right-hand page directs
you to the Internet to help you
find out more. You can log on
to **www.factsforprojects.com**
to find free pictures, additional
information, videos, fun
activities, and further web links.
These are for your own personal
use and should not be copied or
distributed for any commercial
or profit-related purpose.

If you do decide to use the
Internet with your book, here's a
list of what you'll need:
• A PC with Microsoft® Windows®
  XP or later versions, or a
  Macintosh with OS X or later,
  and 512Mb RAM

• A browser such as Microsoft®
  Internet Explorer 8, Firefox 3.X,
  or Safari 4.X
• Connection to the Internet via
  a modem (preferably 56Kbps) or
  a faster Broadband connection
• An account with an Internet
  Service Provider (ISP)
• A sound card for listening to
  sound files

Links won't work?
www.factsforprojects.com is
regularly checked to make sure
the links provide you with lots
of information. Sometimes you
may receive a message saying
that a site is unavailable. If this
happens, just try again later.

Stay safe!
When using the Internet, make
sure you follow these guidelines:
• Ask a parent's or a guardian's
  permission before you log on.
• Never give out your personal
  details, such as your name,
  address, or e-mail.
• If a site asks you to log in or
  register by typing your name or
  e-mail address, speak to your
  parent or guardian first.
• If you do receive an e-mail from
  someone you don't know, tell
  an adult and do not reply to the
  message.
• Never arrange to meet anyone
  you have talked to on the
  Internet.

The publisher is not responsible
for the accuracy or suitability
of the information on any
website other than its own. We
recommend that children are
supervised while on the Internet
and that they do not use Internet
chat rooms.

CONTENTS

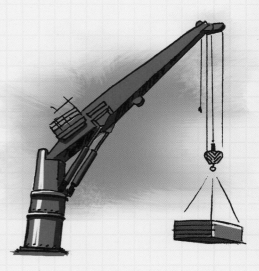

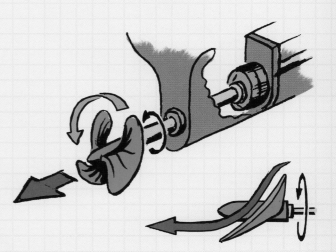

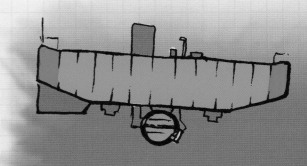

INTRODUCTION

The first people who sat on a floating fallen tree, then tied a few trimmed tree trunks together as a raft, and then hollowed out one log as a canoe started something incredible. Thousands of years before cars and aircraft, early traders transported valuable cargo along rivers and coasts, while explorers crossed oceans to settle new lands. Towns grew up along shorelines, and water became the most important way to travel.

Kayakers use a double-ended paddle to propel, steer, and stop, for complete craft control.

PADDLES AND SAILS

To sail where you want, you need propulsion. Push against water, and it partly pushes back, thus human hands were the first true paddles. Larger surfaces for pushing came with paddles and oars, but they were tiring to use. Sails appeared more than 5,000 years ago to catch wind power. Oars were still needed because the sail design for traveling into the wind, or tacking, did not arrive for another 4,000 years.

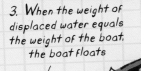

3. When the weight of displaced water equals the weight of the boat, the boat floats

Boat

1. Boat pushes aside or displaces a certain volume of water

Water

2. Displaced water causes an upthrust, force, or buoyancy

SINK OR FLOAT?

The scientific reason why boats float was worked out by Archimedes of ancient Greece. An object lowered into water pushes aside, or displaces, a volume of water. The water pushes back with a force, upthrust, equal to the weight of water displaced. When the object is low enough, the upthrust of the displaced water equals its own weight, and it floats. A heavy stone's weight is always more than the weight of water it displaces, so stones sink.

Tall ships are traditionally rigged sailing vessels that are still very popular today.

>>>SHIPS AND SUBMARINES<<<

The watercraft featured in this book are Internet linked.
Visit www.factsforprojects.com to find out more.

PROPS ARE TOPS

Steam engines that powered the Industrial Revolution in the late 18th and early 19th centuries soon found their way onto ships. The paddlewheels of early steamers were adapted from the waterwheel, which was familiar at the time. Several inventors experimented with designs of propellers, also called water screws, which were a reverse version of the Archimedes screw used for moving water. In 1845, the British Navy raced the paddlewheeler *Alecto* against the screw-driven *Rattler*. The latter won easily, and from that time, props were tops.

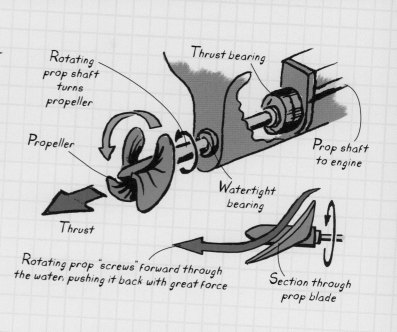

Rotating prop shaft turns propeller

Thrust bearing

Propeller

Prop shaft to engine

Watertight bearing

Thrust

Rotating prop "screws" forward through the water, pushing it back with great force

Section through prop blade

The helmsperson may steer a ship, but the captain is in complete command on the bridge.

Motorcruisers are like traveling personal hotels

BIGGER AND FASTER

Many new kinds of watercraft continued to develop, such as commercial oil supertankers from the 1860s and military aircraft carriers from the 1920s. For fun, we have luxury motorcruisers and offshore powerboats for the rich, and Jet Skis and sailing dinghies for the not-so-rich. Seafaring is easier when a ship's control room, or bridge, bristles with the latest gadgets, such as radar, sonar, satellite weather maps, and GPS satellite navigation.

As fuel prices rise, sail power could return in earnest, but heavy seas, high tides, and unpredictable currents and storms will still challenge the most experienced sailors well into the future.

YACHT

For more than 5,000 years, starting along the Nile River in ancient Egypt, people used the wind to push their ships and boats along. Boats propelled by oars go back farther still, but sailing is easier than rowing, provided there is some wind. Yachts are fairly small, light sailing craft. They are used for fun and pleasure, or for racing, rather than for carrying big loads of cargo.

One of the world's largest sailing yachts is the Maltese Falcon, with masts 190 ft (58 m) tall, 15 sails, and a length of 289 ft (88.1 m)—as long as a soccer field.

Eureka!

Until about 1,100 years ago, sailing ships had mainly square sails that could only be pushed by the wind. Then the lanteen, or triangular sail, was invented, which could be swung around to catch the wind from any angle.

What next?

Modern energy costs are high, but wind power is free. Boat designers are testing big cargo-carrying ships with both engines and sails.

Mast Tall upright poles, or spars, are masts. They hold up the tops of the sails and keep the vertical sides straight.

Jib

✳ How do SAILS work?

The boom holds the sail at an angle to the wind so that the wind pushes it along, partly forward and partly sideways. The boat does not slide sideways because of its hull shape and the large surface area of the keel or centerboard, which both resist sideways movement. The sail also forms a curved shape, like an aircraft wing. Air passes over the curve more quickly than under it (see page 19) to "suck" the boat along.

In 1989, 77-year-old Tawny Pupot sailed across the Atlantic Ocean single-handedly in 72 days.

Deck

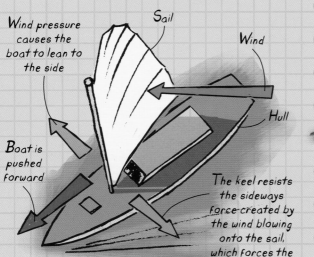

Wind pressure causes the boat to lean to the side

Sail

Wind

Boat is pushed forward

Hull

The keel resists the sideways force created by the wind blowing onto the sail, which forces the boat forward

Keel

Take a guided tour of one the biggest and most luxurious yachts by visiting www.factsforprojects.com and clicking the web link.

Sails Some yachts have one sail, others 20 or more. They are made from very strong tear-proof fabric, either natural fibers such as cotton or flax (linen), or artificial fibers like nylon and polyester.

The ships with the biggest sails were "clippers" bringing tea and spices from East Asia to Europe and North America. The famous Cutty Sark was launched in 1869. It had more than 30 sails with a total area of more than 30,000 sq ft (2,787 sq m)—bigger than ten tennis courts.

Boom The pole or tube forming the boom is attached along the bottom, or foot, of the sail. It can swing the sail around according to the wind direction.

Mainsail

Rigging Various kinds of ropes, or lines, hold up the mast (standing rigging) and move the sails (running rigging). They are made from strong, smooth-running, rot-proof fibers and steel rope.

Windsurfers can reach speeds of 55 mph (89 km/h).

Frames and bulkheads U-shaped frames give the hull (the main body of the boat) its strength and form, and have holes to save weight. Bulkheads are like dividing walls inside the hull.

WINDSURFING

In 1948, Newman Darby attached a sail and a mast mounted on a swivel joint onto a small, flat-bottomed boat and invented the sailboard. The rider holds a double-sided boom and tilts the sail to catch the wind, moving along the board and leaning back so that it does not flip over.

RACING CATAMARAN

The "cat" is a boat with two hulls (main bodies), usually of equal size. It has several advantages over the usual single hull or monohull. For example, it is so wide that it is more difficult to capsize, or tip over onto its side in the water. The hulls, called vakas, are connected by one or more decks, or akas, that hold the masts. Catamarans were used in South Asia for centuries but only became accepted for official racing in the 1970s.

Eureka!

In the 1690s, part-time pirate William Dampier described catamaran-style double canoes in the Bay of Bengal, in the Indian Ocean. However, boat-builders did not start to use the design until Nathanael Herreshoff made a "cat" in 1876/77.

Since the 1970s, huge engine-powered catamarans have become popular as high-speed ferries carrying people, vehicles, and even huge trucks. They are smoother and safer in rough seas than hydrofoils.

Sails "Cats" can have a bigger sail area than a monohull boat because they are more steady and stable in the water.

Boom

Wheel

Rails

✳ How do HULLS AND MULTIHULLS work?

A monohull may tip over if it has heavy loads on deck and tall sails. The deep-keeled monohull is more stable since the keel adds weight low down, and its surface area resists tipping. "Cats" and trimarans (three-hulled craft) have wider bases with a larger deck area for cabins, people, and equipment.

Stern This is the rear or "aft" end of the boat, which usually has the rudder for steering.

Rudder

Monohull   Monohull with deep keel

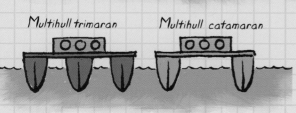
Multihull trimaran   Multihull catamaran

Keel The keel resists sideways pressure and forces the boat to move forward.

Watch an exciting video of a catamaran race by visiting www.factsforprojects.com and clicking the web link.

Alloy mast The mast is made from alloys, or mixtures of metals and other substances, especially aluminum. It is a hollow tube, lighter and stronger than a solid rod.

One of the newest, largest catamarans is Hemisphere, 144 ft (44 m) long and 550 tons (499 metric tons) in weight, with a 171-ft-tall (52 m) mast. It carries 12 guests on luxury cruises around the Caribbean, manned by a crew of eight.

The fastest round-the-world sailing trip was in 2005 by Bruno Peyron in Orange II, taking 50 days 16 hours. He broke the record set by legendary adventurer Steve Fossett and a crew of 12, who took 58 days and nine hours in their catamaran Cheyenne a year before.

A catamaran race, with spinnakers rigged

✳ What are SPINNAKERS?

The spinnaker is a large sail that blows out like a balloon. It works like a normal sail when moving in the direction of the wind, or at an angle downwind. However, unlike a normal sail, the spinnaker cannot sail into the wind. This means that racing sailors put up, or rig, the spinnaker for the downwind parts of their course, then roll or furl it and use the normal sails for the upwind parts of the race.

Bow This is the front or "forward" end of the boat, which is sharp and streamlined to slice through the water easily.

Lightweight hull Modern "cats" have hulls made of very light but strong materials such as fiberglass or carbon-fiber composites.

JET SKI

The PWC, personal watercraft, is often called by the trademarked name Jet Ski. It's a combination of motorcycle and speedboat, powered by a "jet" of water that makes the hull "ski" across the surface. Jet Skis are mainly for fun, racing, and stunt-riding. However they are also useful for emergencies, such as rescuing a swimmer swept away by strong water currents.

Eureka!

The first Jet Skis were invented by bank official and off-road motorbike enthusiast Clayton Jacobson in the early 1970s. They were built by Kawasaki, which also makes fast, powerful motorcycles.

What next?

Some Jet Skis are fitted with a small parachute so that the rider can take off at speed over the crest of a wave, open the parachute to float back down, reel in the parachute quickly, and start all over again.

In 2002, Count Alvaro de Marichalar y Sáenz de Tejada of Spain took four months, riding 12 hours each day, to travel from Rome, Italy, across the Mediterranean Sea to Gibraltar and then across the Atlantic Ocean to Miami, Florida—all on a Jet Ski.

Front fairing

Rub strip

Hull

Engine Some Jet Skis have four-stroke gas engines, others are two-stroke (see page 14). The engine is placed low so the craft is stable and does not tip over.

Handlebars

Engine

Drive shaft

Impeller

Water is sucked into the intake

Ducting

Water is forced out of the nozzle at high speed, pushing the Jet Ski forward

✳ How does WATERJET PROPULSION work?

The impeller sucks in water from below the craft and forces it out rearward at high pressure through the nozzle. The force of the water blasting backward pushes the craft forward. The nozzle is connected by cables to the handlebars. Turning the handlebars to the left makes the nozzle swing to the left, so the craft steers left. The rider can also steer by "leaning" the craft as if on a motorbike.

Discover everything you need to know about personal watercraft by visiting www.factsforprojects.com and clicking the web link.

Hinged handlebars On some Jet Skis the handlebars are part of the main body. On others, they are mounted on a hinged upright tube so the rider can stand up as well as sit down.

Seat

EXPLODED VIEW

Some Jet Skis reach speeds of more than 60 mph (97 km/h), and riders have traveled more than 600 mi (966 km) in 24 hours.

GL·X900

GL-X

In freeriding, Jet Ski riders use waves and surf as jump ramps to take off and reach amazing heights.

A Jet Ski stunt rider gets some "big air"

✳ AQUABATICS!

Jet Ski riders don't just race. They also have freestyle competitions for tricks and stunts such as somersaults, back flips, midair spins and barrel rolls, and even diving under the surface and surging back up again into the air. The riders earn points for each maneuver as well as speed as they race around a track marked by floating buoys.

Drive shaft A long, spinning shaft from the engine turns the impeller. The drive shaft has watertight seals around it so water cannot leak into the engine compartment.

The Jet Ski rider has an ignition key on a rope or line tied to the wrist or life jacket. The key goes into the ignition switch before the engine can start. If the rider falls off, the key comes out and the engine stops.

Impeller Shaped like a fan or propeller, the impeller spins quickly inside its duct (curved tube).

Jet nozzle The jet of water blasts out of a tubelike nozzle to push the Jet Ski along.

Impeller duct Water is sucked in for the impeller through an opening or inlet. This has a meshlike grate or screen to keep out larger objects such as seaweed, driftwood, and fish.

OUTBOARD SPEEDBOAT

A speedboat has a long, streamlined hull with a sharp front or bow. The lower surface is designed to "plane" or skim over the surface at high speed. This is faster than a hull that is partially submerged and tries to push or carve through the water. An outboard engine is one mounted outside the main hull, rather than inboard or within the hull.

Eureka!

The hydroplane is a special design of speedboat devised in the 1950s. It has two sponsons, or short projections, one on each side at the front of the hull. These give stability and lift the craft clear of the water except for its propeller.

What next?

In 1978, Ken Warby reached a speed of 317 miles an hour (510 km/h) in his hydroplane *Spirit of Australia*. Several people have since died trying to break this record.

Until 1911, the world water speed record was held by steam-powered boats.

Controls The steering wheel is linked by cables to the outboard motor and makes it swivel from side to side.

Dash The dash or control board has dials for information such as engine temperature and fuel level.

Fuel tank Fuel for a two-stroke engine has special lubrication oil mixed with the gasoline.

Planing shape The hydroplaning hull is almost flat-bottomed or shaped like a shallow V. It lifts nearly clear of the water at high speed to avoid the water pushing back to cause resistance or drag.

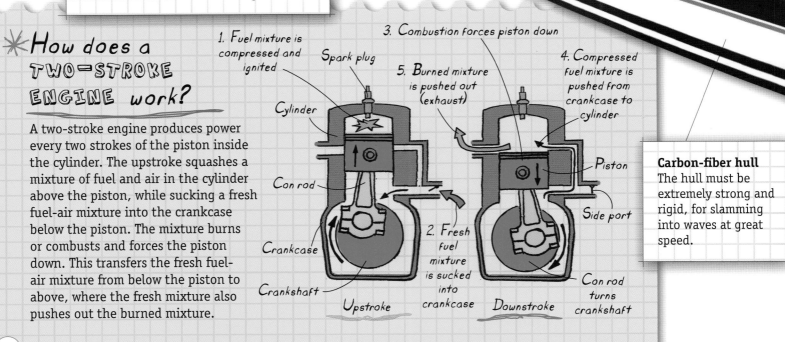

✳ How does a TWO-STROKE ENGINE work?

A two-stroke engine produces power every two strokes of the piston inside the cylinder. The upstroke squashes a mixture of fuel and air in the cylinder above the piston, while sucking a fresh fuel-air mixture into the crankcase below the piston. The mixture burns or combusts and forces the piston down. This transfers the fresh fuel-air mixture from below the piston to above, where the fresh mixture also pushes out the burned mixture.

1. Fuel mixture is compressed and ignited
3. Combustion forces piston down
Spark plug
5. Burned mixture is pushed out (exhaust)
4. Compressed fuel mixture is pushed from crankcase to cylinder
Cylinder
Con rod
Piston
Crankcase
2. Fresh fuel mixture is sucked into crankcase
Side port
Crankshaft
Upstroke
Downstroke
Con rod turns crankshaft

Carbon-fiber hull The hull must be extremely strong and rigid, for slamming into waves at great speed.

Download a video of Ken Warby setting the world speed record by visiting www.factsforprojects.com and clicking the web link.

World water speed ace Ken Warby built his craft, Spirit of Australia, at home. He set the world water speed record in 1978 at Blowering Dam, on the Tumut River in the Snowy Mountains of southeast Australia.

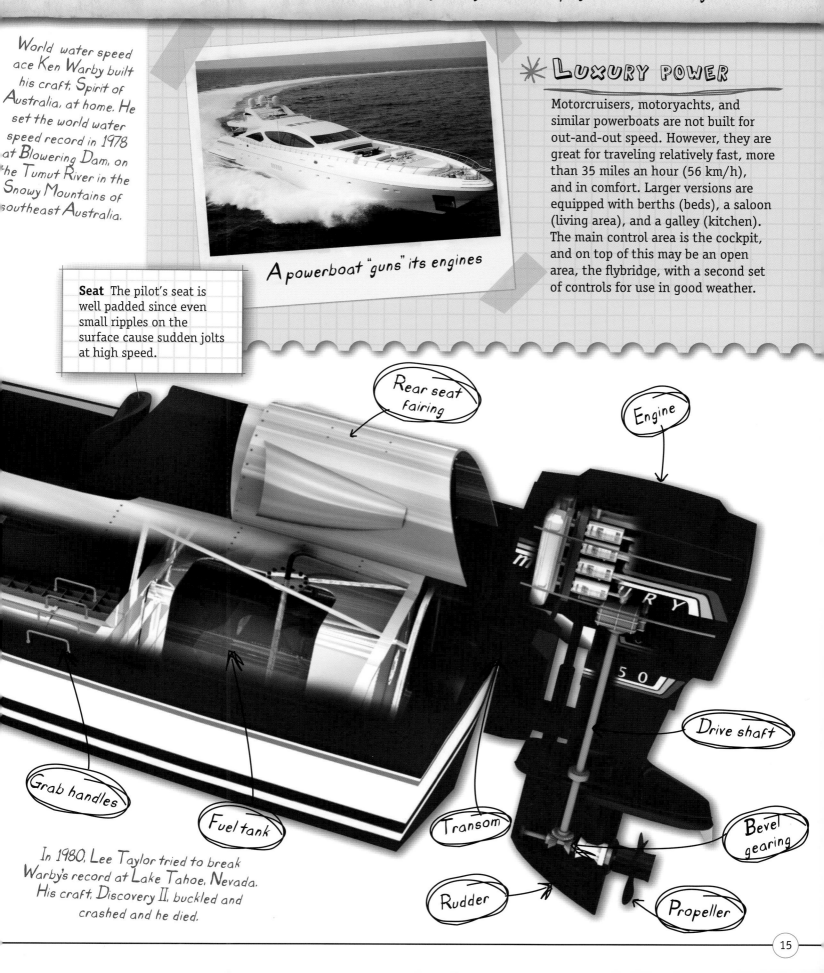

A powerboat "guns" its engines

✳ LUXURY POWER

Motorcruisers, motoryachts, and similar powerboats are not built for out-and-out speed. However, they are great for traveling relatively fast, more than 35 miles an hour (56 km/h), and in comfort. Larger versions are equipped with berths (beds), a saloon (living area), and a galley (kitchen). The main control area is the cockpit, and on top of this may be an open area, the flybridge, with a second set of controls for use in good weather.

Seat The pilot's seat is well padded since even small ripples on the surface cause sudden jolts at high speed.

Rear seat fairing

Engine

Grab handles

Fuel tank

Transom

Drive shaft

Bevel gearing

Rudder

Propeller

In 1980, Lee Taylor tried to break Warby's record at Lake Tahoe, Nevada. His craft, Discovery II, buckled and crashed and he died.

OFFSHORE POWERBOAT

Offshore powerboats are built to withstand waves, wind, and weather away from the shore, out in the open ocean. Slamming into a wave at 100 miles an hour (160 km/h) is almost like hitting a brick wall. Powerboats have to be enormously tough, strong, and rigid. They are usually powered by twin marine diesel or gas engines. In World Championship P1 races, the diesels can be up to 13 liters—six times the size of a family car engine.

Eureka!

The diesel engine was invented by Rudolf Diesel in the early 1890s. It is heavier than a gas engine, but is a useful design where much power is needed, as in powerboats, trucks, and tractors.

What next?

Electric powerboats have been around since 1880. The latest models can travel faster than 60 miles an hour (96 km/h), are quiet, and do not produce polluting fumes.

One of the longest offshore powerboat races is the Round Britain competition, which covers 1,500 mi (2,414 km) at average speeds of more than 60 mph (96 km/h).

Fuel tanks Some powerboats hold well over 200 gallons (757 l) of fuel, enough for cruising for most of the day.

Hull shape Powerboats "plane" or skim with only the propellers and rudders under the surface. The hull is made from aluminum-based metal alloys or carbon-fiber composites.

Rails

Cleat

Inner structure Apart from fuel tanks, inside the watertight hull is mostly air, so the boat is very light. The framework of metal alloy keeps the hull shape rigid.

✳ High speed ACTION

Powerboat racing is extremely expensive—a top team needs to spend tens of millions of dollars. However, it's not just pure speed that wins. The boats must be reliable and economical with fuel. The pilots, or drivers, have to take into account waves, tides, currents, and wind, and keep a constant lookout for dangers such as driftwood. Crashing through the waves is physically demanding, even when strapped into well-padded seats.

An all-out offshore powerboat race

Some powerboat meets consist of several races, such a a sprint of less tha 30 mi (62.8 km) an an endurance of mo than 90 mi (145 km

Read facts, watch videos, and download photos of offshore powerboats by visiting www.factsforprojects.com and clicking the web link.

Air intakes

Airfoil wing

Windshield

Rudder The rudder is linked to the steering wheel by cables running along inside the hull. Some boats have two rudders, one near each propeller.

✳ How do DIESEL ENGINES work?

A diesel engine has four stages or strokes:
1. Inlet: The piston moves down and sucks in a mixture of fuel and air. **2. Compression:** The piston moves up and compresses the mixture.
3. Combustion: The mixture is so hot from compression it explodes, forcing the piston down.
4. Exhaust: The piston goes up and pushes out the burned mixture. The piston's up-down movements turn the engine's main shaft, the crankshaft, by a connecting rod.

Inboard engines The twin marine diesels are at the rear, under large covers that can be removed to service and repair them.

Engine cooling Marine diesels are cooled by seawater taken in continuously through an inlet and expelled through an outlet, rather than water circulating through a radiator, as in a car.

Depending on conditions, some powerboats race at more than 150 mph (241 km/h). However, in some events, if the weather is bad, they are kept to lower speeds by a device called a limiter attached to the engine.

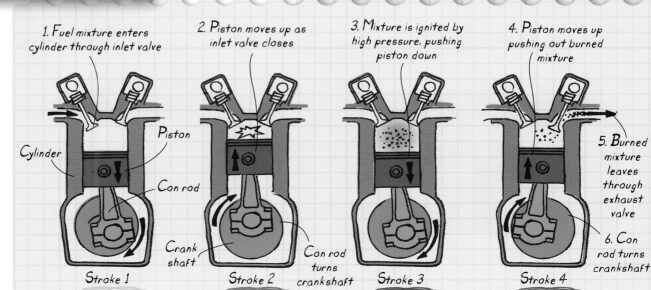

1. Fuel mixture enters cylinder through inlet valve

2. Piston moves up as inlet valve closes

3. Mixture is ignited by high pressure, pushing piston down

4. Piston moves up pushing out burned mixture

Piston

Cylinder

Con rod

Crank shaft

Con rod turns crankshaft

5. Burned mixture leaves through exhaust valve

6. Con rod turns crankshaft

Stroke 1   Stroke 2   Stroke 3   Stroke 4

HYDROFOIL

Hydrofoil watercraft "fly" above the surface on winglike foils attached to struts beneath the hull. The foil works like an aircraft wing to generate a lifting force at high speed, which raises the foil so the craft above it is above the surface. This greatly reduces the drag or resistance of the craft moving through the water. It also means the hull is held clear of rough water and waves.

Eureka!

Hydrofoils were developed in stages by several scientists and engineers, including John Thornycroft around 1900, Alexander Graham Bell (inventor of the telephone) from about 1906, and Enrico Forlanini in 1910.

What next?

Engineers are always working on new foil designs, including "smart" foils that can alter their curved shape according to their speed.

The sit-down hydrofoil is like a chair on a large water ski that "flies" over the water, towed behind a speedboat.

Rails

Bulkheads

Propellers One prop turns faster than the other to steer the craft.

Drive shafts The long drive or propeller shafts extend deep below the boat, so the propellers remain in the water when the hull is lifted clear.

A hydrofoil lifts well over the water

✳ HANDY TRIPS

Commercial hydrofoils run as fast ferries in many areas, especially South and East Asia. They provide a quick smooth ride, handy for trips across lakes, rivers, bays, and sheltered inshore waters. However, they may have trouble with large waves and high winds.

Struts The struts are thin with sharp front and rear edges to produce as little water resistance as possible. On some hydrofoils, they can swivel left or right to work as rudders.

Learn about jetfoils and how they work by visiting www.factsforprojects.com and clicking the web link.

Some surfers have boards fitted with foils to cope with really big waves away from the shore.

Water flow is faster over the foil, resulting in lower pressure

Hydrofoil is sucked upward, supporting the weight of the hull

Strut

Foil shape with curved upper surface

Sea

Trailing edge

Leading edge

Direction of travel

Water flow is slower under the foil, resulting in higher pressure

✳ How do HYDROFOILS work?

A foil's shape is curved when seen edge-on, known as an airfoil (see page 8). The curve is greater on the upper surface than the lower. Water flowing past the foil must go faster over the top than beneath. Faster water means less water pressure. So the lower pressure above the foil sucks it upward with a force called lift, raising the hull out of the water.

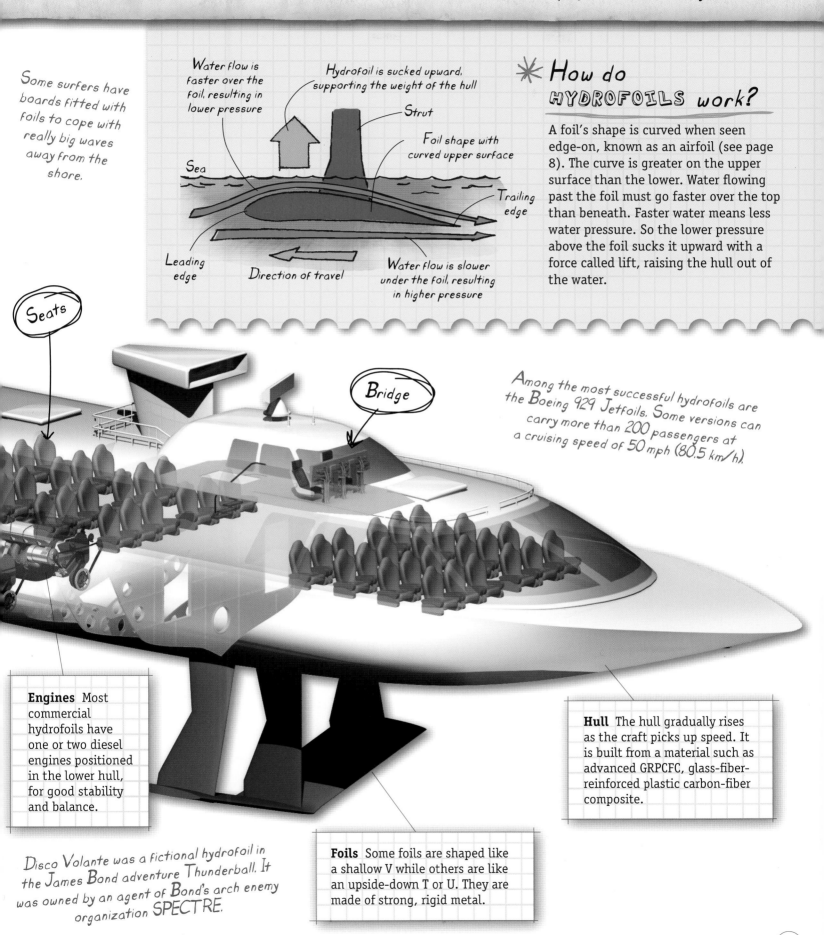

Seats

Bridge

Among the most successful hydrofoils are the Boeing 929 Jetfoils. Some versions can carry more than 200 passengers at a cruising speed of 50 mph (80.5 km/h).

Engines Most commercial hydrofoils have one or two diesel engines positioned in the lower hull, for good stability and balance.

Hull The hull gradually rises as the craft picks up speed. It is built from a material such as advanced GRPCFC, glass-fiber-reinforced plastic carbon-fiber composite.

Disco Volante was a fictional hydrofoil in the James Bond adventure Thunderball. It was owned by an agent of Bond's arch enemy organization SPECTRE.

Foils Some foils are shaped like a shallow V while others are like an upside-down T or U. They are made of strong, rigid metal.

MILITARY HOVERCRAFT

Hovercrafts fly rather than sail, floating just above the surface. They are also known as ACVs, air cushion vehicles. Their great benefit is that they can travel across smooth ground as well as water. The giant LCAC, landing craft air cushion, is the U.S. military's combined boat, low-altitude aircraft, and load-carrier. It can take troops, vehicles, and equipment from a larger transport ship to the shore and even up onto the land.

LCACs carry a load of 75 tons—the weight of an Abrams M1 battle tank.

Eureka!

The first full-size working hovercraft, SRN-1, was built in 1959 by Christopher Cockerell and his team from the Saunders Roe aircraft makers. It hovered at a height of about 10 inches (25.4 cm).

What next?

Personal hovercraft are gaining popularity, but mainly for fun and racing. They are not allowed to travel on roads and are awkward to steer in high winds.

Bow thruster This tube carries some of the air from the lift fans and can be turned to blow it in any direction to help with maneuvering and steering.

Control station

Ramp

✳ How do HOVERCRAFTS work?

A hovercraft floats on a high-pressure cushion of air. Its lift fans (impellers) have large angled blades and are driven by an engine such as a gas turbine (as used in helicopters). The lift fan blows air downward with great speed and force. The air collects in the skirt area below and builds up enough pressure to lift the craft. As the skirt rises, air begins to escape from under its edge, but the pressure remains high enough for the craft to hover.

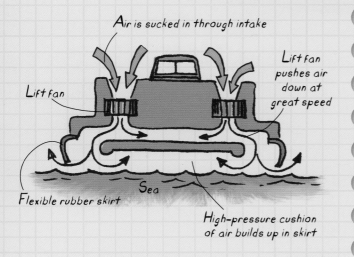

Air is sucked in through intake

Lift fan pushes air down at great speed

Lift fan

Flexible rubber skirt

Sea

High-pressure cushion of air builds up in skirt

The U.S. LCAC is almost 90 ft (27.4 m) long, 50 ft (15.24 m) wide, and 24 ft (7.3 m) high when hovering.

Discover everything you need to know about the amazing LCAC by visiting www.factsforprojects.com and clicking the web link.

LIFT FAN

Bow thruster

Rear lift fan

Front lift fan

Air outlet ducting

The British Griffon 2000 TD hovercraft is in service with more than seven military forces around the world.

Propeller The large aircraft-type propeller is housed in a collar-like shroud with the rudder just behind.

Rudder

Prop shaft Each propeller spins on a shaft driven by a gas turbine engine, which produces more than 4,500 horsepower.

Engine housing

US NAVY

-00

Get the BALANCE right

A hovercraft must be balanced or trimmed so that the weight of the craft plus its load is evenly spread. If not, it might hover nose-down rather than level, or lean to one side and veer around. The loadmaster's job is to put all parts of the load into the correct places and secure them for stable balance.

Outlets

Loading a military hovercraft

Skirt The flexible rubberized skirt hangs down around the hull. It traps air to raise the air pressure and lift the craft so it passes easily over waves.

EXPLODED VIEW

PASSENGER HOVERCRAFT

Hovercraft that carry passengers and vehicles have been used in many parts of the world, although they are less common today. They can fly off the water up onto a ramp to load and unload. They are also fast, cruising at more than 60 mph (96 km/h), and the trip is smooth in calm weather. However, the ride is noisy, and big waves make the craft sway and rock, making some passengers sick.

Eureka!

Early types of hovercraft had no skirts and tipped over easily. The flexible multisection skirt was developed in 1962 by Denys Bliss, a colleague of Christopher Cockerell, and helped rectify the stability problem.

What next?

A new breed of hovercraft are exploring remote swamps. They cause little disturbance to wildlife as they switch off their engines and observe rare animals.

The largest commercial hovercraft included the French N500 Naviplane and the British SR-N4. They weighed about 275 tons (249.5 metric tons), were 160 ft (48.8 m) long, and could carry 400 passengers and more than 50 cars.

The Russian ZUBR are the largest hovercraft in the world, weighing 600 tons (544 metric tons)—165 tons (150 metric tons)more than a fully loaded jumbo jet.

Controls A hovercraft's main controls are the lift throttle or lever to adjust hovering height, the main throttle for speed, and the steering wheel for the rudders.

Bridge The bridge is the control center of a large ship, where the captain and pilot or helmsperson keep watch and steer.

Radar

Seats

Skirt

Life raft containers

✳ How does a HOVERCRAFT STEER?

Like a propeller aircraft, a hovercraft is pushed along by its props. In some versions, the propellers swivel left or right on their pylons to blow air at an angle for steering around corners. Other types have rudders that push the air to the side and make the hovercraft turn. Control is tricky in high winds.

Hovercraft steers left

Propellers push hovercraft forward

Fast-moving air from the propeller pushes past the rudders

Rudder

Rudders deflect air to the left

Watch a hovercraft in action by visiting
www.factsforprojects.com and clicking the web link.

The giant SR-N4 hovercraft went into service in 1968 but
was phased out by 2000 due to increasing fuel costs.

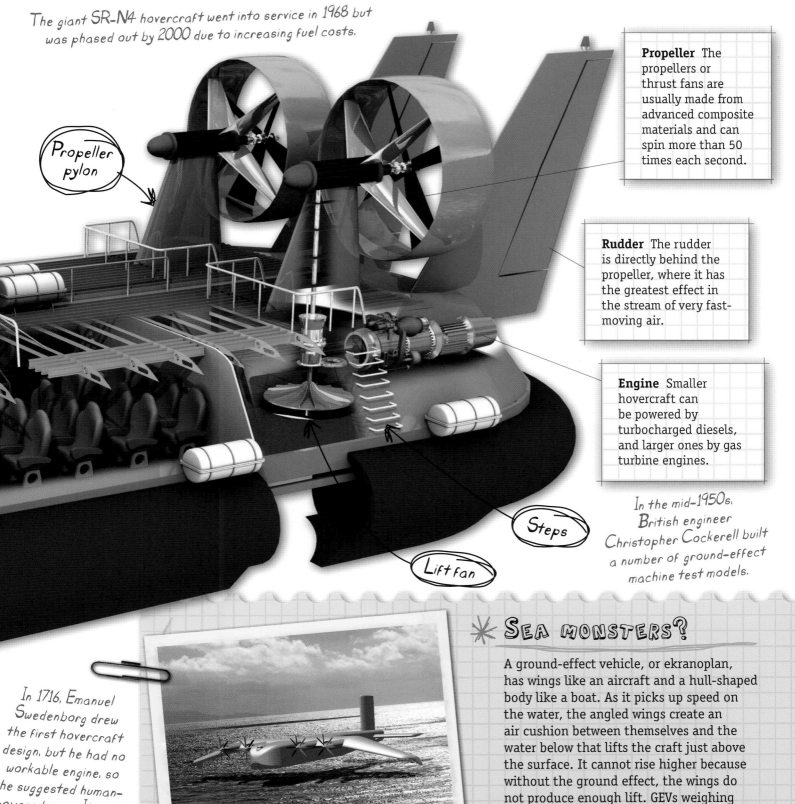

Propeller pylon

Propeller The propellers or thrust fans are usually made from advanced composite materials and can spin more than 50 times each second.

Rudder The rudder is directly behind the propeller, where it has the greatest effect in the stream of very fast-moving air.

Engine Smaller hovercraft can be powered by turbocharged diesels, and larger ones by gas turbine engines.

Steps

Lift fan

In the mid-1950s,
British engineer
Christopher Cockerell built
a number of ground-effect
machine test models.

In 1716, Emanuel
Swedenborg drew
the first hovercraft
design, but he had no
workable engine, so
he suggested human-
powered oars. It would
never have worked!

An artist's impression of the
Boeing GEV

✳ SEA MONSTERS?

A ground-effect vehicle, or ekranoplan, has wings like an aircraft and a hull-shaped body like a boat. As it picks up speed on the water, the angled wings create an air cushion between themselves and the water below that lifts the craft just above the surface. It cannot rise higher because without the ground effect, the wings do not produce enough lift. GEVs weighing more than 550 tons (500 metric tons) and cruising at more than 300 mph (483 km/h), powered by turbojet engines, have been tested for military use on large lakes and inland seas.

CRUISE LINER

It's a luxury hotel with every comfort, from swimming pools to movie theaters and gyms—all traveling smoothly across the ocean. Cruise liners are vacation homes to thousands of passengers as they visit famous ports and sightseeing locations. When a liner docks, passengers go off to visit the sights as the crew work fast to restock the ship with food, water, fuel, and other essentials.

Eureka!

The first cruise liner built for luxury travel was *Prinzessin Victoria Louise*, launched in 1900. It weighed 4,850 tons (4,400 metric tons) and traveled across the Atlantic and Caribbean, as well as the Mediterranean and Black seas. When it ran aground in 1906, the captain felt so guilty that he shot himself.

What next?

Vast cruise ships planned for the future may have a grassy central area like a city park with paths, flower borders, and trees.

Radar mast

One of the largest cruise ships is Queen Mary 2, launched in 2003. At about 165,000 tons (149,685 metric tons), it is 1,128 ft (343.8 m) in length—longer than three soccer fields.

Bridge

Bow

Cargo hold Stores such as fresh water and food are kept toward the bottom of the hull. They must be tied or lashed down to prevent rolling around or falling over in rough seas. Their weight is ballast to increase the ship's stability.

A powerful tug helps a cruise liner into port

✳ Give us a TOW!

It takes several miles for a massive ship to slow down and stop. During this time, it can be blown off course by winds or carried along by tides and currents. Tugs are small, powerful boats that tow big ships using strong cables called hawsers. They may even "nudge" the bigger ships. The tug captain knows local conditions well and inches the ship into the correct position. Two or more tugs may work together for extra control.

Discover everything there is to know about cruise liners by visiting www.factsforprojects.com and clicking the web link.

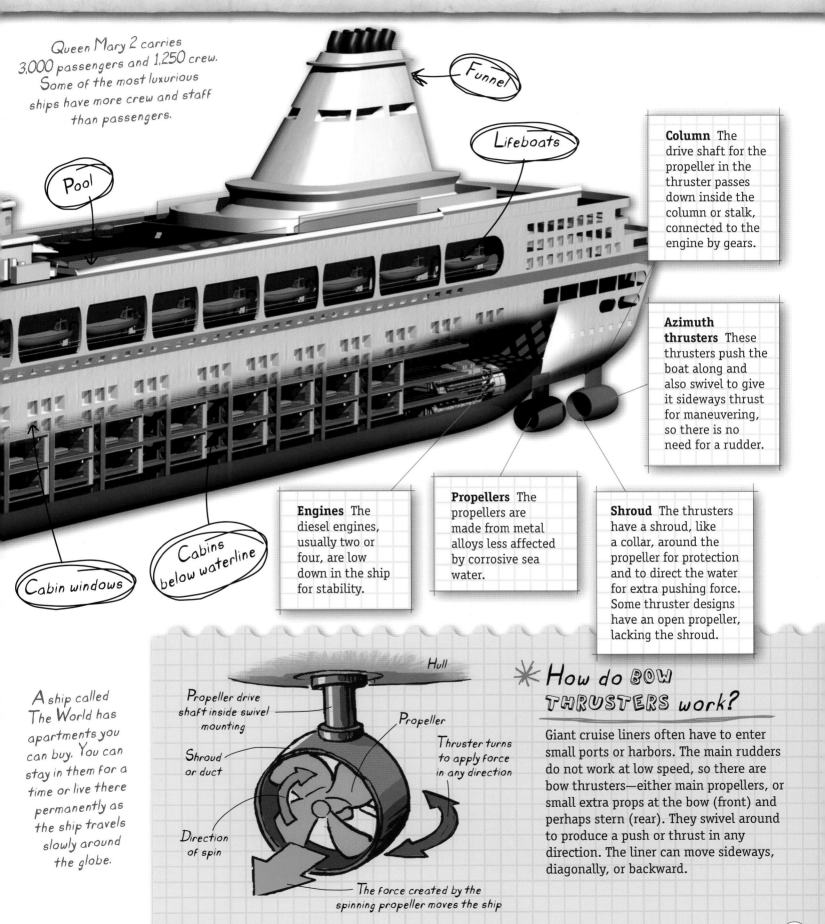

Queen Mary 2 carries 3,000 passengers and 1,250 crew. Some of the most luxurious ships have more crew and staff than passengers.

Funnel

Lifeboats

Pool

Column The drive shaft for the propeller in the thruster passes down inside the column or stalk, connected to the engine by gears.

Azimuth thrusters These thrusters push the boat along and also swivel to give it sideways thrust for maneuvering, so there is no need for a rudder.

Engines The diesel engines, usually two or four, are low down in the ship for stability.

Propellers The propellers are made from metal alloys less affected by corrosive sea water.

Shroud The thrusters have a shroud, like a collar, around the propeller for protection and to direct the water for extra pushing force. Some thruster designs have an open propeller, lacking the shroud.

Cabins below waterline

Cabin windows

Hull

Propeller drive shaft inside swivel mounting

Propeller

Shroud or duct

Thruster turns to apply force in any direction

Direction of spin

The force created by the spinning propeller moves the ship

A ship called The World has apartments you can buy. You can stay in them for a time or live there permanently as the ship travels slowly around the globe.

✳ How do BOW THRUSTERS work?

Giant cruise liners often have to enter small ports or harbors. The main rudders do not work at low speed, so there are bow thrusters—either main propellers, or small extra props at the bow (front) and perhaps stern (rear). They swivel around to produce a push or thrust in any direction. The liner can move sideways, diagonally, or backward.

FREIGHTER

Cargo ships, freighters, and other similar ships are the hard-working "trucks" of the ocean. They crisscross oceans around the world carrying all kinds of freight and cargo, from cars and televisions to food and flowers. As soon as they reach a port, they unload quickly, fill up with another load, and set off once more. Global trade is big business where time is money and cannot be wasted.

Eureka!

Cranes were loading and unloading ships in ancient Greece, more than 2,500 years ago. Mathematician and scientist Archimedes (287–212 BCE) designed one so huge that it could lift enemy ships out of the water!

Mast

Coasters are ships with a shallow hull that does not extend very far down into the water. They can travel through shallow water and over rocks and reefs where ocean-going ships with much deeper hulls would run aground.

Bridge The ship is controlled from the bridge, with the wheel for steering, speed controls, and displays showing information such as engine temperature, fuel levels, and much more.

In 2008, a Japanese cargo ship able to carry 6,400 cars set sail, powered partly by electricity from more than 320 solar panels.

Rudder

Propeller

*How do WATER SCREWS work?

Rotating prop shaft turns propeller

Thrust bearing

Propeller

Prop shaft to engine

Watertight bearing

Thrust

As prop rotates, it "screws" forward, blasting back water with great thrust

Section through prop blade

A water screw or ship's propeller has angled blades that spin around to push the water backward and so thrust the ship forward. They are designed with a foil-type shape so that they are sucked forward as well as pushing backward (see page 19). The drive shaft or prop shaft from the engine passes through several sets of bearings that transfer the force of thrust from the prop to the ship's hull.

Hull Most working cargo ships have hulls made of steel plates welded together for long-lasting strength and toughness. They can also be repaired easily.

Look at some incredible pictures of huge freighters and bulk carriers
by visiting www.factsforprojects.com and clicking the web link.

Hook Different-size hooks can be attached to the cranes, depending on the loads being lifted.

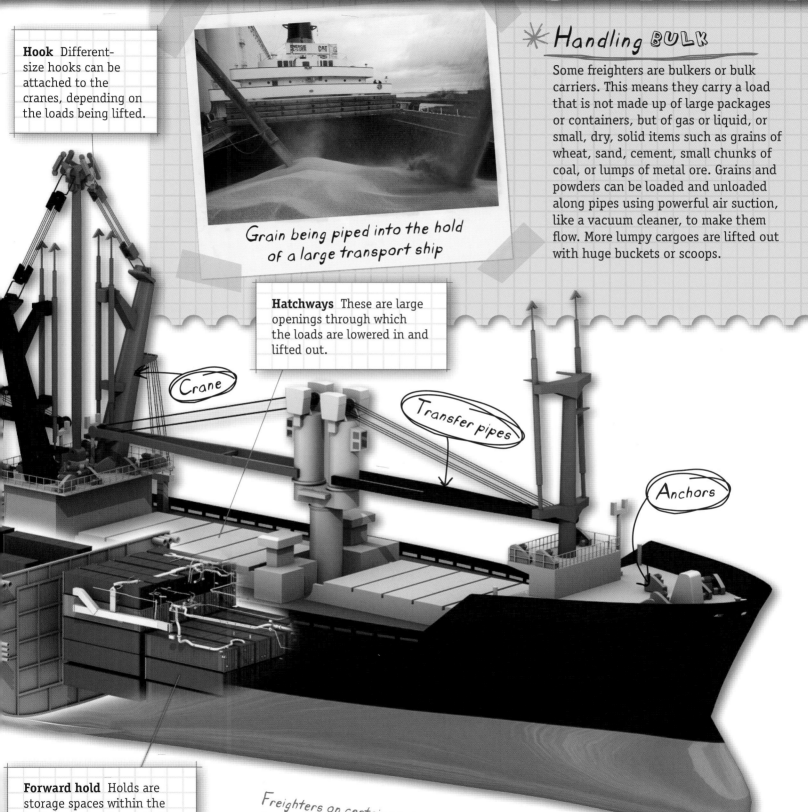

Grain being piped into the hold of a large transport ship

✳ Handling BULK

Some freighters are bulkers or bulk carriers. This means they carry a load that is not made up of large packages or containers, but of gas or liquid, or small, dry, solid items such as grains of wheat, sand, cement, small chunks of coal, or lumps of metal ore. Grains and powders can be loaded and unloaded along pipes using powerful air suction, like a vacuum cleaner, to make them flow. More lumpy cargoes are lifted out with huge buckets or scoops.

Hatchways These are large openings through which the loads are lowered in and lifted out.

Crane

Transfer pipes

Anchors

Forward hold Holds are storage spaces within the hull. Some go almost to the base or keel of the hull; others have separate layers or decks inside.

Freighters on certain routes cannot be too big, otherwise they would not fit into shortcut canals and locks. The width of the Panama Canal's locks in Central America is 110 ft (33.5 m).

Bulbous bow

CONTAINER SHIP

The container ship or "box boat" carries hundreds of containers in its hold and also stacked in piles on its deck. Standard containers are enormous metal boxes, mostly 40 feet (12 m) long by 8 feet (2.4 m) wide by 8.5 feet (2.6 m) high. They are filled with all kinds of goods and materials, and then transported by road, rail, and ship.

Eureka!

Container ships were first used in the 1950s. The earliest ones were made from converted oil tankers that were built during World War II (1939–1945).

What next?

Containers made out of modern composite materials are more expensive to build, but they are much lighter to transport and could pay for themselves in five years.

Bulkheads The strong steel hull has partitions called bulkheads that separate it into watertight compartments. If one is damaged and leaks, the others keep the ship afloat.

Dockside cranes Many container ships only have small cranes. They rely mainly on huge overhanging cranes that run along rails on the dockside to load and unload their containers.

Up to 10,000 containers are probably lost at sea each year. Most slip overboard during high winds or storms, while others sink with their ship.

Winch

✳ How do SHIPBORNE CRANES work?

Many cranes on ships are hydraulic, worked by high-pressure oil. The oil is pumped into a cylinder and pushes along a piston linked by a connecting rod to the crane's "arm"—the jib or boom. The jib lowers to extend the reach of the crane, and the whole crane swivels around on its base. The cable goes around pulleys or sheaves and is reeled in slowly but powerfully by an electric winch.

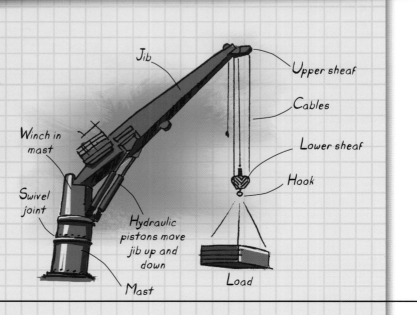

Jib

Winch in mast

Swivel joint

Hydraulic pistons move jib up and down

Mast

Upper sheaf

Cables

Lower sheaf

Hook

Load

Cargo too large to carry in containers can be handled using flat racks, open top containers, and platforms.

Watch amazing videos of container ships in action by visiting
www.factsforprojects.com and clicking the web link.

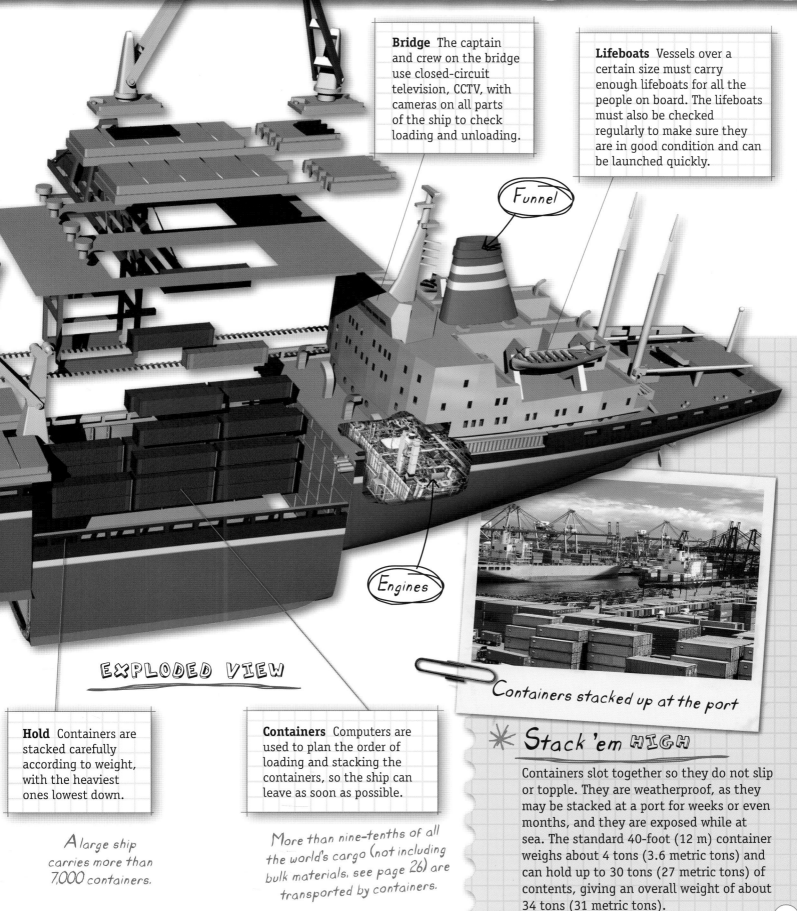

Bridge The captain and crew on the bridge use closed-circuit television, CCTV, with cameras on all parts of the ship to check loading and unloading.

Lifeboats Vessels over a certain size must carry enough lifeboats for all the people on board. The lifeboats must also be checked regularly to make sure they are in good condition and can be launched quickly.

Funnel

Engines

EXPLODED VIEW

Hold Containers are stacked carefully according to weight, with the heaviest ones lowest down.

Containers Computers are used to plan the order of loading and stacking the containers, so the ship can leave as soon as possible.

A large ship carries more than 7,000 containers.

More than nine-tenths of all the world's cargo (not including bulk materials, see page 26) are transported by containers.

Containers stacked up at the port

✳ Stack 'em HIGH

Containers slot together so they do not slip or topple. They are weatherproof, as they may be stacked at a port for weeks or even months, and they are exposed while at sea. The standard 40-foot (12 m) container weighs about 4 tons (3.6 metric tons) and can hold up to 30 tons (27 metric tons) of contents, giving an overall weight of about 34 tons (31 metric tons).

OIL SUPERTANKER

The world's biggest ships are the bulk carriers known as oil tankers. Crude carriers transport crude oil, or petroleum, from oil fields such as those in the Middle East to countries all around the world. Product carriers convey the substances made from crude oil at refineries, such as fuels like gasoline, diesel, and kerosene, as well as lubricating oils, alcohols, and chemical solvents.

Eureka!

Oil tankers were first used in the 1860s in eastern Asia, sailing across the Caspian Sea and along the rivers of northern Europe. By the 1890s, tankers were crossing all major oceans.

What next?

The biggest oil tankers are "sitting duck" targets for pirates, enemies in war, terrorists, and other dangers. Some shipbuilders now use smaller, faster carriers to avoid these risks.

Loading pipes Flexible hoses and rigid pipes carry the pumped oil in and out, and also transfer it between tanks to spread the load for good balance.

Supertankers are more accurately known as ULCCs, ultralarge crude carriers, carrying more than 350,000 tons (317,515 metric tons) of cargo.

A cofferdam is a small space left open between two bulkheads to give protection from heat, fire, or collision. Tankers generally have cofferdams forward and aft of the cargo tanks, and sometimes between individual tanks.

Waterline mark

Hull Most modern tankers are double-hulled, with two layers of "skins" in case one is damaged and leaks. The hard hull surface forms a huge area for reflecting sonar pulses, so other ships can detect the supertanker even at night or in thick fog by sonar as well as radar.

✳ How does SONAR work?

Sonar (sound navigation and ranging) detects nearby objects in water using sound waves. A transmitter sends out pulses or "pings" that travel far and fast through the water. These bounce off objects as echoes that are detected by a receiver. The direction and timing of the echoes shows the position and distance (range) of the object. Sonar is like a sound version of radar (see page 33).

5. Ship's computer displays information

1. Ship emits sound waves or tows sonar probe

2. Probe emits sound waves

4. Echoes (reflected waves) detected by probe

3. Sound waves bounce off surfaces such as seabed

Discover everything you need to know about supertankers by visiting www.factsforprojects.com and clicking the web link.

Pumping control station Huge pumps driven by electric motors force oil into the hold tanks when loading, and out again at the destination. They are usually sited all together in the pump room.

A large oil tanker starts to slow down at least one hour and 12 mi (19.3 km) before it needs to stop. In an emergency, it can stop in about 15 minutes and 2 mi (3.2 km).

Bridge

Crane

Propeller

Cofferdam

Long hull

Oil tanks As the oil is unloaded and the tanks empty, they are filled with inert gases that do not combine or react with other substances. Otherwise the oil fumes might catch fire and explode.

Bulkheads One vast tank inside the hull would allow oil to slosh around and upset the ship's balance. So bulkhead walls divide it into many compartments.

By weight, oil tankers form more than one-third of all cargo carried at sea.

The world's largest supertanker was Seawise Giant, built in 1979. It was extended, damaged by enemy attack, altered again, and had names including Happy Giant and Jahre Viking. Finally, it became the 1,500-ft-long (457 m) floating oil storage tank Knock Nevis, moored off Qatar in the Middle East.

Oil pollutes the sea from the oil tanker Braer off the Scottish coast in 1993

✳ Environmental DISASTERS

Oil tanker accidents have caused some of the greatest environmental disasters of modern times. The oil leaks out as a floating slick that smothers the water's surface and kills fish, seabirds, seals, corals, and other marine life. One of the worst spills was in 1991 when the ABT Summer leaked more than 275,000 tons (249,476 metric tons) of oil off Angola, Africa.

AIRCRAFT CARRIER

Some of the largest ships in the world are floating air bases designed to go to anywhere for war or, hopefully, to guard the peace. In addition to carrying jet aircraft and helicopters, these giant vessels have other weapons, such as guided missiles, torpedoes, and mines. They also act as command, control, and communications centers for other seacraft in their fleet, such as battleships and destroyers.

Eureka!

In 1910, Eugene Ely was the first pilot to take off from a ship, the armored cruiser USS *Birmingham*. The next year, he was first to land on a ship, the USS *Pennsylvania*, another cruiser with a specially built platform.

What next?

Stealth planes scatter radar signals with their angled surfaces, edged shapes, and radio-absorbing paint. Stealth ships with similar features have also been tested and will become more common.

Nimitz class carriers have two nuclear reactors powering four steam turbines and can cruise at more than 30 mph (48 km/h).

✳ How do carriers store AIRCRAFT?

Carriers have huge flat elevators—platforms that rise up and down between the lower decks and flight deck, transporting aircraft and equipment. Nimitz carriers have four of these along the sides, called upper-stage elevators. They only go down two decks. Then their loads have to be moved sideways to lower-stage elevators inside the hull, which take them down farther.

Fuel tanks Huge tanks store jet fuel for the aircraft. Many carriers are nuclear powered so their plutonium fuel pellets could fit into a small truck.

Flight deck Longer than three soccer fields, the flight deck is about 250 feet (76.2 m) wide at its broadest part.

Catapult rail

Air defense guns

Anchor

Hull The steel alloy hull on a Nimitz class carrier is nearly 1,000 feet (304.8 m) long, with a flight deck about 66 feet (20 m) above the water.

Island

Flight deck

Crane

Hangar entrance

Aircraft have folding wings to take up less space

Upper stage elevator forms part of flight deck when raised

The flight deck of an aircraft carrier is one of the world's most dangerous places to work.

Discover lots more information about aircraft carriers by visiting www.factsforprojects.com and clicking the web link.

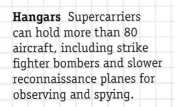

Bridge The bridge (ship's control room), aircraft control center, radar, and radio equipment are in the "island" on the right or starboard side, leaving the main deck free for planes.

Radar

Aircraft landing

Elevators

68

Workshops Servicing and repair must be done on the aircraft while at sea, with a full range of parts and spares.

Key carrier crew members include the "shooter" who oversees the steam-powered aircraft launch catapults, and the handler, who controls where aircraft are positioned and stored so that they are available as needed.

Hangars Supercarriers can hold more than 80 aircraft, including strike fighter bombers and slower reconnaissance planes for observing and spying.

✳ The importance of RADAR

Radar (radio direction and ranging) works like sonar (see page 30) but uses radio waves instead of sound. Radar can detect nearby ships, shorelines, and hazards such as icebergs showing above the surface, as well as aircraft or other objects in the sky many hundreds of miles away. Radar cannot be used beneath the surface because radio waves travel very poorly underwater.

Big carriers have a ship's crew of up to 3,000 people, and another 3,000 aircraft crew including pilots and maintenance engineers.

A complex array of radar and radio masts on a modern carrier

SUBMARINE

A submarine is designed to travel underwater for long distances. (Submersibles go deeper but not for long distances, see page 36.) Most large subs are military craft, known as the "silent service" because they can remain hidden under the surface and undetected for weeks or even months. To do this, they must carry all their fuel, food, and other stores, although they can make fresh water to drink and also vital oxygen to breathe, both from sea water.

Eureka!

The first submarines were built in the 1620s by Cornelius Drebbel. Powered by oars, they were rowed along the Thames River to impress King James I. In 1776, David Bushnell built the *Turtle*, the first military submarine, to attack British ships in the Revolutionary War (1775–1783).

What next?

One of the strangest sports is submarine racing. Spectators try to spot the positions of the submarines by the bubbles they release or by their periscopes—if they can catch a glimpse!

Ballistic missile

Periscope The periscope is a vertical telescope that extends from the tower when the sub is "below" (submerged), to see around at the surface.

Tower The fin, sail, or conning (control) tower is where people enter and leave through hatches, and where the crew can keep watch when the sub is at the surface.

Bridge

✳ UNDERSEA WORLD

Many coastal areas now have tourist submarines where visitors can travel beneath the surface to watch fish, corals, and other underwater life. It's easier than scuba diving! The latest versions are 60 feet (18.3 m) long, carry 50 passengers down to 300 feet (91.4 m), and stay under for more than an hour. They are powered by electric motors driven by rechargeable batteries so they are quiet and cause almost no pollution.

Bunks

Tourists enjoy an all-around view

Sonar sphere

In 1958, the U.S. submarine Nautilus sailed past the North Pole under the floating raft of ice that covers most of the Arctic Ocean.

EXPLODED VIEW

Carry out an experiment to discover how submarines dive and rise by visiting www.factsforprojects.com and clicking the web link.

Most submarines can only descend a few hundred yards at the most. American Seawolf subs would probably be crushed below 1,600 ft (487.7 m). The Russian Komsomlets sub could reach 4,000 ft (1.2 km).

Galley

Mess room

Nuclear reactor The nuclear reactor is contained in a radiation-proof casing. In many subs, it only needs refueling every 12–15 years.

Turbines Heat from the nuclear reactor boils water that pushes at huge pressure past the angled blades of a turbine, making it spin to turn the propellers.

Propeller The prop is designed to work very quietly, without forming too many swirls or bubbles, so that other boats cannot hear or see the sub approaching.

726

Planes Hydroplanes tilt to make the submarine rise or fall, and the rudders steer it left or right.

Ballast tanks These chambers are between the outer waterproof hull and the inner pressure hull that withstands the incredible pressure at depth.

Some submarines can stay underwater for more than six months. The limit is not so much fuel, drinking water, or oxygen, but food for the crew.

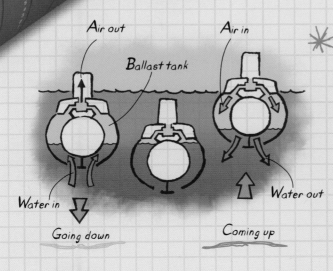

Air out

Ballast tank

Air in

Water in

Going down

Water out

Coming up

☀ How do subs WORK?

A sub goes up or down using the ballast tanks between its inner and outer hulls. To sink, it allows air out of the tanks and water in. This makes the sub heavier than the water around it, so it descends. To rise, compressed air made from seawater is forced at high pressure into the tanks. Here the air expands, and although it weighs the same as when compressed, it takes up more space—which was formerly occupied by much heavier water. The sub is now lighter, so it ascends.

DEEP-SEA SUBMERSIBLE

Submersibles are specialized to dive very deep but not travel too far (unlike submarines). They are used mainly for scientific research, surveys, and exploring wrecks. Crewed submersibles must have air inside, and the pressure of the water around would crush an ordinary submarine like paper. The best shape to withstand all-around pressure is round, like the sphere of *Trieste*. In 1960, it descended to Earth's lowest point, the Challenger Deep in the northwest Pacific. No craft or people have ever been nearer to the center of the Earth.

Eureka!

In 1985 in the northwest Atlantic Ocean the remote submersible *Argo*, a sledlike design with cameras, located the wreck of the massive liner *Titanic*. It had sunk in 1912 and rested 12,500 feet down on the seabed.

What next?

An "undersea town" where people live and work will allow scientists to study ocean wildlife. They would not have to keep coming up to the surface, which involves slow depressurization to avoid the dangerous condition known as the "bends."

Piccard and Walsh spent 20 minutes on the Challenger Deep seabed, eating chocolate for extra energy.

ROVs (remotely operated vehicles) are uncrewed submersibles controlled by a long cable, the umbilical or tether. Some have TV cameras and mechanical hands so the operator at the surface can see and hold items.

Propeller

Water ballast tanks
The water tanks were positioned at the front and rear of the upper float part.

Gasoline tanks
Filling most of the upper float part, these tanks contained 3,000 cubic feet (84,951 l) of gasoline.

✳ How did TRIESTE work?

The crew sphere had to withstand pressures of more than one ton on a fingernail-size area. Made of 5-inch (12.7 cm) thick steel, it weighed 8.8 tons (8 metric tons) in water and would sink like a stone. So it was made buoyant by the upper float part containing lighter-than-water gasoline. The water ballast tanks were adjusted so the craft descended slowly. At the end of the dive, release magnets allowed iron ballast to fall out of the ballast hoppers, making the whole craft lighter so it could rise.

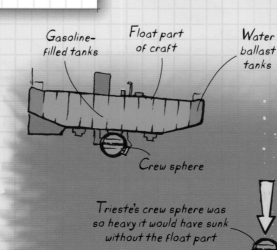

Gasoline-filled tanks

Float part of craft

Water ballast tanks

Crew sphere

Trieste's crew sphere was so heavy it would have sunk without the float part

Seabed

Ballast hopper
Nearly 10 tons (9 metric tons) of ballast or weight in these two chambers was in the form of small iron pellets.

Release magnets

Discover deep-sea machines, how they work, and what they find by visiting www.factsforprojects.com and clicking the web link.

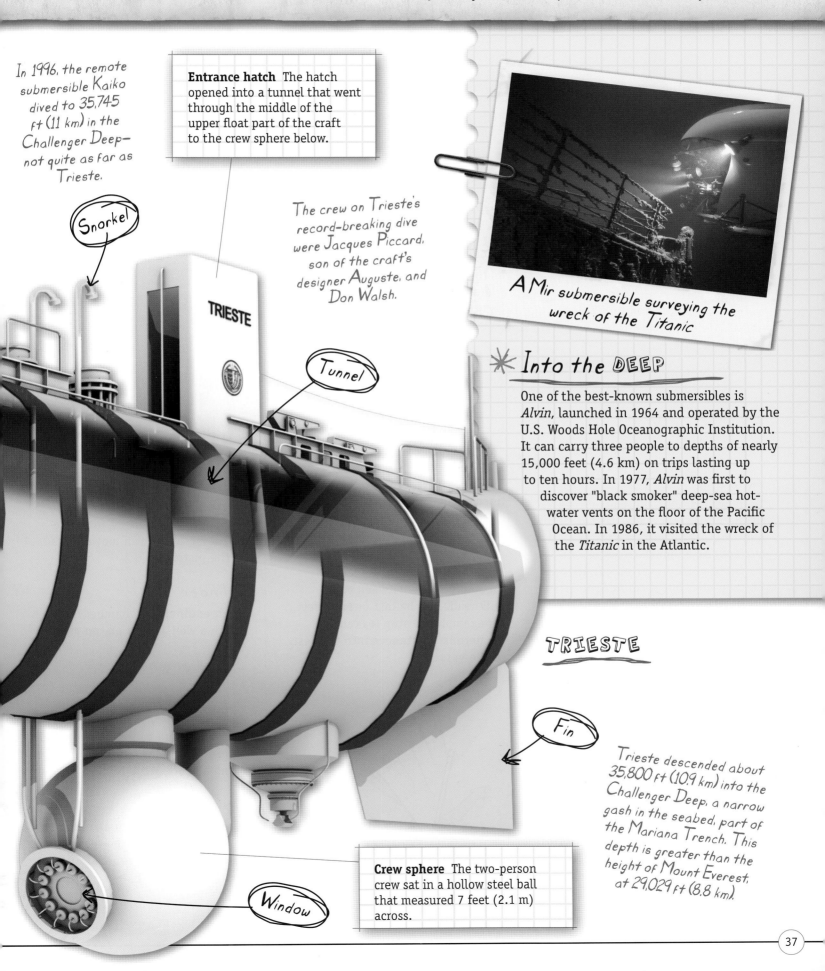

In 1996, the remote submersible Kaiko dived to 35,745 ft (11 km) in the Challenger Deep—not quite as far as Trieste.

Entrance hatch The hatch opened into a tunnel that went through the middle of the upper float part of the craft to the crew sphere below.

Snorkel

The crew on Trieste's record-breaking dive were Jacques Piccard, son of the craft's designer Auguste, and Don Walsh.

TRIESTE

Tunnel

A Mir submersible surveying the wreck of the Titanic

✳ Into the DEEP

One of the best-known submersibles is *Alvin*, launched in 1964 and operated by the U.S. Woods Hole Oceanographic Institution. It can carry three people to depths of nearly 15,000 feet (4.6 km) on trips lasting up to ten hours. In 1977, *Alvin* was first to discover "black smoker" deep-sea hot-water vents on the floor of the Pacific Ocean. In 1986, it visited the wreck of the *Titanic* in the Atlantic.

TRIESTE

Fin

Trieste descended about 35,800 ft (10.9 km) into the Challenger Deep, a narrow gash in the seabed, part of the Mariana Trench. This depth is greater than the height of Mount Everest, at 29,029 ft (8.8 km).

Crew sphere The two-person crew sat in a hollow steel ball that measured 7 feet (2.1 m) across.

Window

GLOSSARY

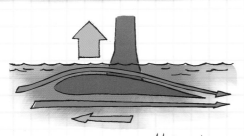

Hydrofoil

Airfoil

The cross-section shape of most aircraft wings, being more curved or humped on the upper surface than the lower surface, to provide a lifting force. Also the cross-section shape for various water vessel parts such as propellers (water screws) and the foils of hydrofoils.

Alloy

A combination of metals, or metals and other substances, for special purposes such as great strength, extreme lightness, resistance to high temperatures, or all of these.

Azimuth

To do with angles and movements in the horizontal plane, that is, left and right, level with the water's surface. An azimuth thruster can swivel to point in various directions horizontally, but it cannot tilt vertically to point up or down.

Ballast

Heavy substances, such as water, concrete, or metal, added to a craft or vehicle to make it more stable and to keep it from toppling over, for example, in high winds or when turning at a high speed. Many boats and ships have ballast tanks low in the hull that can be flooded with water if the vessel is not loaded heavily enough, so that they float at the correct level and remain steady while moving and turning.

Boom

In watercraft, the pole or spar along the base, or foot, of a sail. It prevents the bottom area of the sail from flapping out of control.

Bow

In watercraft, the front or pointed end of the hull, or main body.

Bridge

In watercraft, the control room of a large boat or ship, housing the wheel (steering wheel), engine throttles, instrument displays, and other important equipment.

Bulkhead

An upright wall or partition across the width of a structure, such as across the hull of a ship or the fuselage of an aircraft.

Cleat

A device or structure for attaching lines or ropes, which may be shaped like a cylinder, a "T," a "V," or a similar design.

Cofferdam

An empty space between two bulkheads, for ballast or for safety, for example, to stop leakage of dangerous substances into the rest of the vessel.

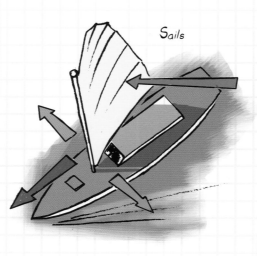

Sails

Con rod

Connecting rod, an engine part that links the piston to the main crankshaft.

Crankcase

The main body of a gasoline or diesel engine, below the pistons and cylinders, containing the crankshaft and con rods.

Thruster

Crankshaft

The main turning shaft in an engine, which is made to rotate by the up-and-down movements of the pistons.

Cylinder

In an engine, the chamber inside which a well-fitting piston moves.

Diesel engine

An internal combustion engine (one that burns or combusts fuel inside a chamber, the cylinder), which uses diesel fuel and causes it to explode by pressure alone rather than by using a spark plug.

Drive shaft

A spinning shaft from an engine or motor that drives or powers another part, such as the propeller of a water vessel.

Gasoline engine

An internal combustion engine (one that burns or combusts fuel inside a chamber, the cylinder) that uses gasoline and causes this to explode using a spark plug.

Hold

In larger ships and boats, the place where cargo, freight, and supplies are kept.

Hull

The main body or central part of a water vessel.

Hydrofoil

A watercraft that uses winglike foils to create a lifting force as they move forward, and so make the hull rise, either partly or completely above the surface.

Impeller

A propeller, rotor, or spinning fan inside a tube, collar, or shroud, to take in a fluid at one end and force it out at the other end.

Jib

A smallish, usually triangular sail in front of the mainsail(s), toward the front of the vessel. This is also the term used for the working arm of a crane.

Keel

A large flap or flange sticking down below the center of the hull, with its flat surfaces facing to each side. Keels are used mainly in sailing vessels to give greater control and stability. The term is also used for the central part of a vessel's structure, like its "backbone" running from front to rear.

Hovercraft

Mast

An upright or almost upright pole or spar that holds up and supports one or more sails.

Nozzle

A cone- or trumpet-shaped device from which fast-moving liquid or gas comes out, as in a Jet Ski.

Piston

A wide, rod-shaped part, similar in shape to a food can, that moves along or up and down inside a close-fitting chamber, the cylinder.

Planing (hydroplaning)

In watercraft, moving by skimming over the surface with almost no part in the water, rather than by pushing through it.

Propeller

A spinning device with angled blades, like a rotating fan, which turns to draw in fluid such as water or air at the front, and thrust it powerfully backward.

Rigging

Lines (as the ropes are called on ships), masts, sails, and other equipment that make a sailing ship travel along under wind power.

Rudder

The control surface of a craft, usually a flat part sticking down below the rear of a boat or ship. It angles to the side to make the craft steer left or right.

Shroud

A collar- or tubelike structure around a spinning impeller (propellerlike fan), which restricts and controls the direction and force of the fluid passing through it.

Stern

In watercraft, the rear or blunt end of the hull, or main body.

Transom

The flat rear panel across some watercraft, which is usually almost vertical and faces backward.

Throttle

A control that allows more fuel and air into an engine for greater speed, sometimes called an accelerator.

Thruster

A small propeller, nozzle, or jet-like part that squirts or puffs out liquid or gas, usually to make small adjustments to a craft's position or its direction of travel.

Turbine

A set of angled fanlike blades on a spinning shaft, used in many areas of engineering, from pumps and cars to boat and jet engines.

Jet Ski propulsion

INDEX